Things to Know About Solar Energy Storage

First published by Kjøller 2023

Copyright © 2023 by Kjøller

Disclaimer:

The information contained in this book is provided for general informational purposes only. While every effort has been made to ensure that the information is accurate and up-to-date, The Author makes no representations or warranties of any kind, express or implied, about the completeness, accuracy, reliability, suitability, or availability with respect to the information, products, services, or related graphics contained in the book for any purpose.

The Author disclaims any liability for any loss or damage, including without limitation, indirect or consequential loss or damage, or any loss or damage whatsoever arising from loss of data or profits arising out of, or in connection with, the use of this book.

Readers are solely responsible for determining the appropriateness of the information contained in this book for their specific purposes and should seek professional advice before acting upon any information contained herein. The Author shall not be liable for any damages of any kind arising from the use of this book or the information contained herein.

Table of Contents

Introduction

Solar energy is rapidly gaining popularity as a sustainable and efficient source of power. As more and more people seek to harness the power of the sun for their daily energy needs, it is essential to understand the technology behind it. One of the most critical components of solar energy systems is energy storage.

Energy storage plays a vital role in balancing the supply and demand of energy, particularly in situations where energy consumption is high or in places where sunlight is not available consistently. It involves collecting and storing excess energy that would otherwise be lost and releasing it when needed. As such, it is crucial to understand the terminologies associated with this process to make informed decisions.

This book aims to provide a comprehensive resource for anyone interested in solar energy storage. It is a glossary type containing easy-to-understand definitions of technical terms and concepts related to energy storage. Whether you're a professional in the energy sector, a student, or an individual seeking to invest in sustainable energy solutions, this book will equip you with the knowledge needed to navigate the world of solar energy storage.

AC Coupling

AC coupling is a method of connecting a battery storage system to an existing photovoltaic system via an inverter that only converts direct current (DC) into alternating current (AC) when it is necessary to store excess energy. When the battery is discharged, the inverter switches the battery voltage to AC, which allows it to charge up again.

Active Cooling

A type of battery cooling system that uses fans, pumps or other external technologies to regulate temperature. Active cooling is a vital component of battery management systems, as it can help prevent overheating and extend the life of the battery. However, active cooling systems require a constant power source, which can reduce overall efficiency.

Ampere-hour

Ampere-hour (Ah) is a unit of electrical charge that measures how much electrical energy a battery can store. It is defined as the amount of electric charge transferred by a current of one ampere for one hour.

Amp-Hour (Ah)

A unit of measurement used to quantify the capacity of a battery. An amp-hour is equal to the rate of current flow (in amps) over time (in hours). A battery with a rating of 100Ah can provide 1 amp of current flow for 100 hours, 10 amps for 10 hours, or any other combination that equals 100Ah.

Autonomous Energy System

This refers to an off-grid system that is designed to be completely self-sufficient and independent from the electrical grid. It utilizes solar energy storage to supply power for all or a portion of the energy needs of a home, business or facility.

Battery Management System

A battery management system (BMS) is a system that monitors and controls a battery pack's state of charge, cell voltage, and temperature. The BMS ensures that the battery operates at maximum efficiency and helps prevent overcharging or undercharging, which can damage the battery's cells.

Capacity

Capacity refers to the total amount of electrical energy that a battery can store. It is usually measured in ampere-hours (Ah) or kilowatt-hours (kWh).

Charge Controller

A charge controller is a device that regulates the flow of electricity between the solar panel and the battery to prevent overcharging or undercharging of the battery.

Cycle Life

Cycle life is the number of times that a battery can be charged and discharged before it starts to degrade significantly or fails. The cycle life of a battery is affected by several factors, including depth of discharge, temperature and the rate of charging and discharging.

DC Coupling

DC coupling is a method of connecting a battery storage system directly to a photovoltaic system. The battery charges and discharges as the photovoltaic system produces and consumes electrical energy. This method does not require an inverter, which can increase efficiency and reduce costs.

Depth of Discharge

Depth of discharge (DOD) is the amount of the total capacity of a battery that has been discharged. It is usually expressed as a percentage. For example, if a battery has a capacity of 100 Ampere-hours and it has discharged 50 Ampere-hours, the DOD is 50%.

Discharge Rate

Discharge rate is the rate at which electrical energy is released from a battery. It is typically measured in amperes (A) or kilowatts (kW). The discharge rate of a battery can affect its cycle life and should be taken into consideration when designing a solar energy storage system.

Economic Benefits

Solar energy storage systems can provide economic benefits in the form of energy savings, increased property value, and potential revenue generation through energy arbitrage, excess energy resale, and demand response programs.

Efficiency

The measure of how well a solar energy storage system converts solar energy into usable electricity. Efficiency is typically expressed as a percentage, with higher efficiency indicating a more productive system. High-efficiency solar panels can be used to store more energy efficiently by absorbing more sunlight.

Electric Grid Resilience

A battery storage system equipped with solar panels can act as backup power in case of a power outage, providing energy security and resilience to the electric grid.

Electric Vehicle Charging

A battery storage system can charge electric vehicles when solar power production is high, which not only reduces the carbon footprint of transportation but also maximizes the utilization of solar energy.

Electric Vehicles (EVs)

Vehicles powered by an electric motor and a battery, which can be charged using solar energy storage systems. EVs are becoming more popular as they are a clean and efficient alternative to conventional gasoline-fueled vehicles.

Electrolyte

A liquid or gel that facilitates the movement of ions between the positive and negative electrodes in a battery. Electrolytes are essential for the proper functioning of a battery, allowing electric current to flow between the anode and cathode.

Emissions Reduction

A solar energy storage system can reduce greenhouse gas emissions by decreasing reliance on fossil fuels in energy production and transportation.

Encapsulation

A technique used to protect the components of a solar energy storage system from environmental factors such as moisture, heat, and mechanical stress. Encapsulation involves creating a barrier or coating around the components to prevent damage, increase durability, and prolong the life of the equipment.

End-of-Life

The stage of the life cycle of a solar energy storage system where the equipment is retired or recycled. End-of-life management involves the proper disposal of equipment components, minimizing environmental impact, and complying with regulatory requirements.

Energy Arbitrage

The practice of buying electricity from the grid when prices are low and storing it in batteries, and then selling it back to the grid when prices are high, which can help reduce energy costs.

Energy Cycles

The number of charge and discharge cycles a battery can go through before its capacity begins to degrade. Higher energy cycles mean a longer lifespan for a battery and better performance for the solar energy storage system. Proper maintenance and monitoring can help maximize energy cycles and extend the life of a battery.

Energy Demand Management

The ability to manage and reduce energy consumption during peak periods can help avoid blackouts, brownouts, and reduce strain on the electric grid, ultimately improving the reliability and affordability of energy.

Energy Density

A measure of how much energy can be stored in a given volume or mass of a storage medium. Higher energy density means more energy can be stored per unit volume, which is an important factor in selecting the right battery for a solar energy storage system. Batteries with higher energy density can store more energy in a smaller space.

Energy Independence

Battery storage systems coupled with solar panels can provide homeowners and businesses with energy independence, freeing them from dependence on the grid and reducing the risk of power outages.

Energy Management System

A software-based system that integrates with solar panels to optimize power production and minimize energy waste. It may manage battery storage, charge controllers, and inverters to maintain the optimal state of the system. The system can also manage and monitor the energy consumption and communicate with the grid to set tariffs.

Energy Management System (EMS)

A software-based tool that helps control and manage the various components of a solar energy storage system. EMS can detect the level of energy storage in batteries and can direct the flow of energy from different sources accordingly. It can also communicate with the utility grid to balance energy demand and supply.

Energy Monitoring

The ability to monitor energy production, utilization, and storage can help identify areas of improvement and optimize energy usage, maximizing energy savings and return on investment.

Energy Payback

The amount of time it takes for a solar panel or battery storage system to produce enough energy to offset the energy used in its manufacturing, transportation, and installation. An efficient solar cell can have a shorter energy payback period, which means it can deliver more carbon-free power over its lifetime.

Energy Storage System

A device that stores excess solar energy for use when solar production is low or when energy demand is high. Energy storage systems can be composed of different types of batteries, ranging from lead-acid to lithium-ion. The type of battery used depends on the required energy density, durability, and cost.

Energy Transformation

The process of converting solar energy into electricity, and storing it in batteries for later use. The transformation can occur through photovoltaic cells, which convert sunlight into direct current, and through inverters, which convert DC into AC power.

Environmental impact

The effect of a solar energy storage system on the environment during its life cycle, which includes manufacturing, transportation, installation, use, and end-of-life. Environmental impact can be reduced by selecting the most environmentally-friendly materials, increasing energy efficiency, and recycling.

Environmentally Friendly

Solar energy is a clean, renewable source of power that does not produce any greenhouse gas emissions or air pollution. Battery storage systems coupled with solar panels further reduce reliance on fossil fuels, making it an environmentally friendly solution.

Expansion

The addition of energy storage systems or components to an existing solar energy storage system. Expansion may involve adding additional panels, batteries, or inverters to the existing system or upgrading the existing equipment to improve system performance.

Fault current

The current that flows when an electrical fault (e.g. short circuit) occurs in a solar energy storage system. Fault current can cause damage to equipment and poses an electrocution hazard to personnel. Proper protection, such as fuses, circuit breakers, and grounding systems, must be installed to prevent fault current from reaching dangerous levels.

Fire safety

Fire safety measures must be taken into account when installing solar energy storage systems. Lithium-ion batteries, commonly used in these systems, can be susceptible to thermal runaway, a condition in which they overheat and catch fire. Proper ventilation, fire suppression systems, and other safety measures can minimize fire risk.

Fire-retardant coatings

Coatings that can be applied to the battery cells in a solar energy storage system to minimize the risk of fire. These coatings are typically made of non-flammable materials that absorb or block heat, preventing thermal runaway from occurring.

Float charge

The maintenance charge applied to a battery in a solar energy storage system to keep it fully charged. Float charge voltage is lower than the battery's maximum charge voltage and is typically applied when the battery is not in use, such as during periods of low solar energy production.

Flow battery

A type of rechargeable battery that generates electricity through the electrochemical reactions of two liquid electrolytes separated by a membrane. Flow batteries are commonly used in solar energy storage systems, as the electrolytes can be stored separately and recharged independently, allowing for flexible capacity and longer discharge times.

Flywheel energy storage

This technology uses a spinning rotor to store energy as kinetic (motion) energy. When electricity is needed, the rotor's energy is converted back to electrical energy. Flywheel energy storage systems are highly efficient and can discharge power quickly, making them useful for certain grid applications.

Frequency regulation

The process of adjusting the power output of a solar energy storage system to maintain a steady frequency in the electricity grid. Frequency regulation is essential to maintain grid stability, as rapid changes in frequency can damage grid equipment and cause power outages.

Fuel cell

A device that generates electricity through a chemical reaction between hydrogen and oxygen. Fuel cells can be used in conjunction with solar energy storage systems to provide backup power or to increase overall energy efficiency by converting excess solar energy into hydrogen for later use.

Full cycle

A full charge and discharge cycle of a battery in a solar energy storage system. Each full cycle reduces the battery's overall capacity and degrades its performance over time. Minimizing the number of full charge-discharge cycles can prolong the battery's useful life.

Fuse

A protective device used to prevent overcurrent and fault current in a solar energy storage system. Fuses are designed to break the circuit when the current exceeds a certain level, preventing damage to equipment and reducing the risk of fire or electrocution.

Heat Loss

Energy storage systems like batteries and capacitors have a limited lifespan, mainly because they lose efficiency over time. Heat loss is one of the factors that contribute to this loss of efficiency, as it increases the rate of chemical reactions within a battery, leading to the depletion of active materials and reduced storage capacity. Proper temperature control and management is, therefore, a key factor in maintaining the longevity of batteries in energy storage systems.

High Voltage Battery

A high voltage battery is a type of rechargeable battery used in energy storage systems. These batteries are designed to handle high loads and are ideal for large scale energy storage, making them suited for commercial and industrial applications.

High-Capacity Battery

A high-capacity battery is a type of rechargeable battery that's able to store larger amounts of energy than traditionally used batteries of the same size. These batteries are commonly used in energy storage systems since they offer greater energy density and can provide backup power for longer periods of time.

Home Battery

A battery storage system that stores electricity generated by solar panels, allowing homeowners to use the energy at a later time. A home battery system is usually connected to a home's electrical panel, cutting down on energy bills and reducing the dependence of the grid.

Hot Weather Performance

Hot temperatures can negatively affect the performance of batteries in energy storage systems. This is because high temperatures produce more internal resistance in batteries, reducing the amount of electrical energy they can store. Hot weather performance is an important consideration when choosing a battery storage system because it has a direct impact on its efficiency and lifespan.

Hybrid Inverter

A hybrid inverter is an electronic device that manages the flow of electricity between solar panels, battery storage, and the electricity grid. Unlike regular inverters, hybrid inverters can switch between on-grid and off-grid modes, and can also operate both AC and DC output voltages, which makes them more efficient for solar energy storage.

Hybrid Solar Systems

These are energy systems that combine solar panels with other sources of energy such as wind turbines or conventional generators. They are designed to be more reliable and efficient as they create a backup system, allowing power to be generated even when there is less sunlight available.

Hydroelectric Power

This type of power is generated by using the energy from moving water to power turbines, which convert the kinetic energy of the water into electricity. Hydroelectric power can be used to store and generate energy, particularly during peak energy usage times when the electricity grid is under stress.

Hydrogen Energy Storage

This is the process of using electricity to break down water into hydrogen and oxygen, which can then be stored and used as an energy source. Hydrogen energy storage is particularly beneficial for long-term storage solutions because it's able to store large amounts of energy, is clean and versatile, and can be used as a transportation fuel.

Hysteresis Control

Hysteresis control is a technique used to manage the flow of electrical energy in energy storage systems. The method involves setting upper and lower voltage thresholds, which allow the battery to store a specific amount of energy and discharge it when the voltage drops below the lower threshold. This helps to prolong the lifespan of the battery by preventing it from being overcharged or undercharged. Hysteresis control is an important aspect of ensuring that energy storage systems operate efficiently and reliably.

Interconnection

This refers to the process of connecting a solar energy system to the electrical grid. Interconnection involves the installation of a bi-directional electric meter which allows excess energy generated from the solar panels to be fed back into the grid.

Inverter

This is an electronic device that converts the direct current (DC) power generated from solar panels to alternating current (AC) power which is used by appliances in homes or businesses. Inverters are essential for solar arrays to function properly and efficiently.

J-Box

Also called a junction box, this device is used to connect the solar panels in parallel or series configurations. It is usually located close to the panels and contains fuses, circuit breakers, or diodes to protect them.

Jigawatt (JW)

A unit used to measure power generated or consumed in a system. Often associated with the movie "Back to the Future," where the time machine required 1.21 jigawatts of power.

Jolt

A sudden burst of energy released from a battery or solar panel. Joules are converted to a flow of energy called jolts when used.

Joule

A unit used to measure the amount of energy released or consumed in a system. In solar energy storage, joules are used to measure the amount of energy stored in batteries. The higher the number of joules, the more energy the battery can store.

Joule Heating

The heat produced by a flow of electrical current through a conductor. In solar energy storage systems, joule heating can be a significant concern because it can lead to battery damage or failure. A well-designed system will include measures to control joule heating and prevent overheating.

Joule Thief

A circuit used to boost the voltage of a low-power input to a higher level. Often used to power small electronic devices or LED lights with a low power source, such as a single AA battery.

Juice

A slang term for energy or electricity. Solar energy storage is often referred to as having "juice" or "power" available for use.

Jumper

A device used to connect two terminals or batteries. It is used to increase the energy storage capacity of a battery array.

Jumper Cable

A cable used to connect two batteries in parallel. It is used to increase the storage capacity of the batteries and provide a backup energy source. Jumper cables are also used to jump-start vehicles from a dead battery.

Junction Box

A device used to connect the solar panels in series or parallel configurations. It contains diodes that prevent reverse current flow and protect the solar panels from damage.

Kapton tape

A specialized tape used in solar panel installations to protect the solar cells from damage and ensure maximum efficiency.

Kelvin (K)

A unit of measure for temperature. It is important in solar energy storage systems to monitor the temperature of the batteries to ensure optimal performance and longevity.

Kilovolt-ampere hour (kVAh)

A unit of measure for the total energy consumed or generated in an AC electrical system over a period of time. It is used to calculate the efficiency of a solar energy storage system.

Kilowatt (kW)

A unit of measure for power output. It represents the amount of power that is generated or used at any given point in time. Solar energy storage systems are rated by their kW output.

Kilowatt Hour (kWh)

A unit of energy commonly used to measure the energy storage capacity of batteries for solar power systems. It represents the amount of energy that can be used by a device that consumes 1 kW of power for one hour.

Kilowatt-hour (kWh)

A unit of measure for energy storage capacity. It represents the amount of energy that can be stored or used in one hour at a rate of 1,000 watts. The higher the kWh rating of the solar battery, the more energy it can store.

Kinetic energy storage

A type of energy storage where energy is stored in the motion of a moving object such as a flywheel. It is used as a short-term energy storage solution.

kVA

A unit of measure for the apparent power in an AC electrical system. It is used to calculate the size of the inverter needed for a solar energy storage system.

KWh Benefit-Cost Analysis

This term refers to the process of evaluating the economic benefits and costs of installing a solar energy storage system. It considers factors such as the cost of the system, the energy savings, and the impact on the environment.

KWh Charge Controller

This term refers to the device that regulates the amount of energy that is fed into the battery from the solar panels. It is important to ensure that the battery is charged properly and that it is not overcharged or undercharged.

KWh Charge Rate

This term refers to the rate at which energy can be charged into the battery. It is an important factor to consider when selecting a solar energy storage system since it can determine how quickly the battery can be charged.

KWh cost

The cost of one kWh of energy saved or used. The cost of kWh varies depending on the location, time of day, and electricity provider. The lower the cost of kWh, the more cost-effective the solar energy storage system is.

KWh cycle life

The total number of charge and discharge cycles a solar battery can go through before its capacity degrades to a certain point. A higher kWh cycle life means that the battery can be used for a longer period of time before needing replacement.

KWh Depth of Discharge

This term refers to the percentage of the battery's total capacity that can be used before it needs to be recharged. A higher kWh depth of discharge means that the battery can be used for a longer time.

KWh Discharge Rate

This term refers to the rate at which energy can be discharged from the battery. It is an important factor to consider when selecting a solar energy storage system since it can determine how quickly the battery can be used up.

KWh Inverter

This term refers to the device that converts the DC power generated by the solar panels into AC power that can be used in the home. It is an important component of the solar energy storage system.

KWh Life Cycle Cost

This term refers to the total cost of ownership of a battery system over its lifespan. It includes the initial cost of the battery, maintenance costs, and replacement costs.

KWh Roundtrip Efficiency

This term refers to the efficiency of the battery to convert energy during the charging and discharging process. A higher kWh roundtrip efficiency means that the battery can store and use more energy effectively.

KWh Stand-alone System

This term refers to a solar energy storage system that is not connected to the grid. It is used in areas where there is no access to the grid and can provide reliable energy for homes and businesses.

KWh Storage Capacity

This term refers to the amount of energy that can be stored in a battery for use during hours of no sunlight. A higher kWh storage capacity means that the battery can store more energy and can be used for a longer time.

KWh Throughput

This term refers to the amount of energy that can be discharged from a battery during its lifetime. A higher kWh throughput means that the battery can be used for a longer time.

KWh/Day

This term refers to the average amount of energy generated by a solar system in one day from the sun. It is used to estimate the capacity of a solar system based on the energy needs of a household.

kWp

A unit of measure for the maximum power output of a solar panel under optimal conditions. It stands for kilowatts peak and is used to rate the power output of solar panels.

Lead-acid battery

An older technology commonly used in off-grid solar installations due to their low cost and reliability. They are less efficient and have a shorter lifespan compared to lithium-ion batteries.

Levelized cost of energy (LCOE)

A metric used to compare the cost of different renewable energy sources or energy storage systems over their entire lifespan, taking into account the initial installation costs and ongoing maintenance and operating costs.

Life cycle assessment (LCA)

A methodology used to evaluate the environmental impact of a product or service over its entire lifespan, taking into account factors such as resource extraction, manufacturing processes, and disposal.

Lithium iron phosphate (LiFePO4) battery

A newer type of lithium-ion battery that is becoming increasingly popular in solar energy storage systems due to its high energy density, long lifespan, and safety features. LiFePO4 batteries are less prone to overheating and have a lower risk of thermal runaway compared to other lithium-ion batteries.

Lithium-ion battery

These batteries are commonly used in solar energy storage systems due to their high energy density, long lifespan, and low maintenance requirements. They are efficient and reliable, making them a popular choice for residential and commercial solar power systems.

Load profile

The pattern of energy consumption throughout the day or week. An understanding of the load profile is essential for designing a solar energy storage system that can provide energy during times of high demand.

Load shifting

A technique used to optimize solar energy storage, where excess solar energy is stored during the day and used when energy demand is high, such as during peak hours. Load shifting can reduce the reliance on the electrical grid during times of high demand, and save money on electricity bills.

Longevity

A key consideration when investing in a solar energy storage system. Systems with a longer lifespan can provide better returns on investment in the long term.

Maximum Power Point Tracking (MPPT)

A technology used in solar charge controllers to maximize the output of solar panels by continuously tracking the point at which the input voltage and output current are at their maximum levels. This ensures that the solar system is operating at its most efficient level.

Megawatt-hour (MWh)

A unit of energy used to measure the amount of electricity produced or consumed over a period of time. One megawatt-hour is equivalent to one million watt-hours.

Microinverters

Devices used to convert direct current (DC) generated by solar panels into alternating current (AC) used in homes and businesses. Unlike traditional string inverters, microinverters operate on a per-panel basis, allowing for better efficiency and optimization of energy production.

Monitoring

The process of regularly monitoring the performance of a solar energy system, including energy production and storage, as well as any potential issues or concerns that may arise. Monitoring is important for ensuring the ongoing efficiency and effectiveness of the system.

Mounting system

The hardware used to attach solar panels to a roof or other surface. Mounting systems can be either fixed or adjustable, and are designed to securely hold panels in place while allowing for maximum energy production.

Nano-crystal technology

A type of solar panel technology that uses small crystals to absorb and convert sunlight into electricity. This technology has the potential to increase the efficiency of solar panels and produce more energy from the same amount of sunlight.

National Electrical Code (NEC)

A set of standards and regulations created by the National Fire Protection Association to ensure safe electrical installation and maintenance practices. These regulations must be followed when installing solar panels and associated equipment.

National Grid

A company responsible for the transmission and distribution of electricity in the United States, including the integration of renewable energy sources like solar. National Grid is actively investing in grid modernization to better accommodate the growth of solar and other renewables.

Net energy metering

A billing arrangement where excess solar energy generated is sent back to the grid and credited to the homeowner's or business's electric bill. This allows for more efficient use of solar energy and can save money on energy costs.

Net metering

A billing arrangement where excess solar energy generated is sent back to the grid and credited to the homeowner's or business's electric bill. This allows for more efficient use of solar energy and can save money on energy costs.

Net-zero energy

A building or home that produces as much energy as it consumes. This is achieved through the installation of solar panels and other energy-saving technologies, making the building more energy-efficient.

Nickel-cadmium (NiCad) batteries

Rechargeable batteries that can be used for solar energy storage. They are known for their high level of durability and reliability, as well as their ability to operate in extreme temperatures.

Non-lithium batteries

Alternative battery options for solar energy storage, such as lead-acid and flow batteries. Non-lithium batteries are more affordable, but typically have a shorter lifespan and are less efficient than their lithium counterparts.

NREL

The National Renewable Energy Laboratory, a government-funded research facility focused on renewable energy and energy efficiency. NREL conducts research on solar energy and provides technical assistance to various stakeholders.

Off-grid

Off-grid refers to a solar energy storage system that is not connected to the electrical grid. This type of system is typically used in remote areas where connecting to the grid is not feasible or cost-effective. Off-grid solar systems store excess energy in batteries to provide power during times when the sun is not shining. These systems are becoming more popular as the cost of solar panels and batteries decreases and the technology becomes more efficient.

Off-grid inverters

Off-grid inverters convert the DC power generated by solar panels and stored in batteries into AC power that can be used to power household appliances. Off-grid inverters are designed to work with a wide range of battery chemistries and configurations and are a critical component of off-grid solar energy storage systems.

Off-peak

Off-peak refers to periods of time when electricity demand is lower than usual. Solar energy storage systems can be used to store excess energy generated during peak hours and use it during off-peak hours when electricity is cheaper.

Operating temperature

Operating temperature refers to the range of temperatures in which a battery can operate efficiently. Solar energy storage systems are designed to operate at specific temperature ranges to optimize their efficiency and lifespan. Battery manufacturers typically provide information on the recommended operating temperature range for their products.

Optimization

Optimization refers to the specific configuration of a solar energy storage system to maximize its efficiency and performance. This may include selecting the right battery technology, orientation of solar panels, and implementing various energy management strategies.

Overcharge

Overcharging is a common issue with solar energy storage systems. If a battery is overcharged, it can become damaged and may need to be replaced. In order to prevent overcharging, charge controllers are used to regulate the amount of energy going into the batteries. These controllers monitor the battery's charge level and adjust the charge rate to prevent overcharging.

Over-discharge

Over-discharging a battery can cause permanent damage to the battery and reduce its lifespan. This can occur when a battery is discharged beyond its recommended depth of discharge. To prevent over-discharging, charge controllers are used to monitor the battery's state of charge and disconnect the load when the battery is nearing the recommended depth of discharge.

Parallel Connection

Parallel connection is a method used to connect multiple photovoltaic panels or batteries together in a solar energy system. This connection allows the panels or batteries to work together in a more efficient way, maximizing the amount of energy that can be generated.

Passive Solar Design

Passive solar design is an architectural approach that incorporates elements that take advantage of the sun's energy to provide heating and cooling in a building. This includes features like large windows, thermal mass, and shading devices that are used to maximize solar gain and minimize heat loss.

Peak Demand

Peak demand refers to the maximum amount of electricity that is required by a particular utility or grid at a given time. This demand usually occurs during periods of high energy consumption, such as during the summer months when air conditioning use is at its highest.

Photovoltaic Efficiency

Photovoltaic efficiency refers to the amount of solar energy that is converted into usable electricity by a photovoltaic panel. The efficiency of a panel is one of the most important factors to consider when designing a solar energy system, as it directly impacts the amount of energy that can be generated.

Photovoltaic Module

A photovoltaic module is a basic unit of a solar panel that consists of a group of photovoltaic cells. The module is designed to capture sunlight and convert it into usable electricity, which is then passed through a power inverter and distributed to the places where it is needed. The efficiency of a module is an important factor to consider when designing a solar energy system, as it directly impacts the amount of energy that can be generated.

Photovoltaic Panels

Photovoltaic panels, also known as solar panels, are devices that capture the energy from sunlight and convert it into usable electricity. These panels consist of arrays of photovoltaic cells that are designed to absorb sunlight and release electrons, which generate a current of electricity. When sunlight hits the panel, the electrons are excited, creating an electrical current that goes to an inverter for conversion to high voltage AC power.

Power Distribution

Power distribution refers to the process of transmitting electricity generated by a solar energy system to the places where it is needed. This can include distribution through power lines or transmission via other methods such as wireless networks.

Power Inverter

A power inverter is a device that converts the DC electricity generated by a solar panel into AC electricity that can be used to power household appliances and other devices. This device is an essential part of any solar energy system, as it allows the energy generated by the panels to be used in a practical way.

Power Storage

Power storage refers to the ability of a solar energy system to store the energy that it generates. This allows the system to be used at times when there is no sunlight available, such as during nighttime or periods of cloudy weather. Power storage can be accomplished through the use of batteries or other storage technologies, which allow the energy to be stored for later use.

PV Switching

PV switching refers to the process of switching between different energy sources in a solar energy system. This can involve switching between solar panels and battery storage or between solar panels and a grid connection, depending on the needs of the system.

Quality of Service

The level of reliability and predictability of a battery storage system's performance. This refers to how consistently the system can deliver power to the grid or to a household, for instance. Higher quality service can lead to more reliable and cost-effective energy storage solutions.

Quantity of Solar Panels

The number of solar panels used to generate energy. This can impact energy storage solutions as systems with larger solar panel arrays can generate more energy to be stored in batteries.

Quantum Dots

Nanoparticles that can be used to improve the efficiency of solar panels. By using quantum dots, solar panels can more easily absorb specific wavelengths of light, increasing their overall efficiency and power output.

Quasi-Fermi Level

A measure of the energy level of electrons in a semiconductor. In solar panels, quasi-Fermi level differences can be used to measure the voltage and efficiency of the panel.

Quasi-Static Simulation

A type of simulation used to analyze the performance of battery storage systems. Quasi-static simulations can model the flow of power in and out of the system over time, allowing for optimization and fine-tuning of a storage system.

Quenching

A process of cooling down a battery to reduce the chances of overheating or failure. This is a critical safety measure for battery storage systems, as overheated or damaged batteries can be highly dangerous.

Quick Charge

A high-speed charging process for battery systems. Quick charging can be useful for electric vehicles and other applications where fast charging is critical, as it can reduce downtime and increase productivity.

Quiescent Current

The small, continuous flow of electricity in a battery storage system, even when it is not actively being used. This can be problematic as it can reduce the overall efficiency of the system over time.

Quiet Time

The period of the day or night when energy usage is low, and solar panels may be producing more energy than is being used. This is an ideal time for battery storage systems to charge up, storing energy for when it is needed most.

Quotas

Government-implemented targets or requirements for renewable energy sources. Some countries or regions may require a certain percentage of energy to come from renewable sources like solar, which can incentivize energy storage and other renewable energy solutions.

Recycling

Solar batteries, like any other electronic product, have a limited lifespan. When they are no longer useful, they can be recycled for their valuable components such as lead, lithium, and nickel. Recycling batteries reduces the need for new materials, conserves resources, and minimizes environmental harm.

Redistribution of power

A solar energy storage system can store excess solar energy generated during the day and redistribute it in the evening when energy use is high. This process can take advantage of the lower cost of daytime solar energy and avoid peak electricity rates in the evening, saving money on energy bills.

Redox flow batteries

A type of rechargeable battery that stores energy in chemical solutions. Redox flow batteries are characterized by their long lifespan, scalability, and ability to store large amounts of energy, making them feasible for commercial and industrial applications.

Remote monitoring

A solar energy storage system can be monitored remotely, allowing users to keep track of their home's energy use, solar generation, and battery status via a smartphone app or web portal. This type of monitoring can help users optimize energy consumption, reduce waste, and enhance the overall performance of the solar system.

Renewable energy

Solar energy is a type of renewable energy generated by converting sunlight into electricity. Unlike fossil fuels that are finite and deplete over time, renewable energy sources are endless and can be replenished continuously. Renewable energy systems are environmentally friendly, sustainable, and contribute to reducing carbon emissions.

Renewable energy credits

Renewable energy credits (RECs) are financial incentives offered by governments to encourage the use of renewable energy sources. Owners of renewable energy systems can receive credits for the energy their systems generate, which can be sold to utilities or other organizations to help meet renewable energy requirements.

Residential energy storage

Residential energy storage involves installing solar batteries within homes to store excess solar energy generated during the day for use at night or during blackouts. This technology provides homeowners with more control over their energy usage and can offer cost savings.

Resilience

The ability of a solar energy storage system to maintain power supply during a blackout or power outage is known as resilience. Solar batteries can store energy for use during emergencies, making them an important factor in creating resilient homes and communities that can continue functioning even during natural disasters or grid failures.

Return on investment (ROI)

ROI measures the efficiency of a solar energy storage system. It calculates the time it takes for the system to pay for itself through energy savings. Factors that can affect ROI include initial system cost, tax incentives, local electricity rates, and usage patterns.

Round-trip efficiency

This is the measure of how much energy is lost during charging and discharging of a solar battery. A high round-trip efficiency means that more of the energy stored in the battery is available for use, while a low round-trip efficiency indicates that significant energy is lost during the process.

Self-consumption

Self-consumption is a term used to describe the use of solar-generated electricity for your own consumption rather than sending it back to the grid. It allows homeowners and businesses to save money on their energy bills by using solar energy instead of buying it from the grid. Self-consumption can be achieved by using a solar energy storage system or by using solar energy during the day when the sun is shining.

Silicon

Silicon is a chemical element commonly used in solar panels to convert sunlight into electricity. It is used in photovoltaic cells because of its ability to absorb sunlight and release electrons, which generates electricity. Silicon is the second most abundant element in the Earth's crust after oxygen.

Smart Inverter

A smart inverter is an inverter that can communicate with other devices in a solar PV system. It allows for better management of the system's energy and can increase its efficiency. A smart inverter can also connect to the internet, allowing for remote monitoring and control.

Solar Battery

A solar battery is a rechargeable battery that is used to store the energy produced by solar panels. It can store excess energy during peak sunlight hours and release it during off-peak hours when demand is high or when there is no sunlight. They are commonly used in off-grid solar applications and are becoming more popular in grid-tied systems.

Solar Charge Controller

A solar charge controller is a device that regulates the voltage produced by a solar panel to prevent overcharging of the battery. It is used to protect the battery from being damaged due to overcharging or undercharging.

Solar Generator

A solar generator is a portable device that uses solar energy to generate electricity. It typically consists of a solar panel, battery and inverter. They are commonly used in off-grid applications such as camping or outdoor events where electricity is needed but not available.

Solar Panel

A solar panel is a device that converts sunlight into electricity. They are made up of several smaller units called photovoltaic cells, which convert the energy in sunlight into electrical energy. Solar panels are generally used for solar power generation and can be installed on rooftops or on the ground. The amount of electricity produced by a solar panel depends on its size, the efficiency of the cells, and the amount of sunlight it receives.

Stand-alone System

A stand-alone solar system is a system that is not connected to the grid. It typically consists of a solar panel, battery, and inverter, and is commonly used in remote areas where access to electricity is limited or nonexistent. Stand-alone systems are also used for backup power in case of power outages.

Storage System

A solar energy storage system is a device that stores excess electricity generated by a solar panel system for use at a later time. The storage system can be used during times when the sun is not shining or during periods of high demand. They are commonly used in residential and commercial solar applications to maximize the use of solar-generated electricity.

System Efficiency

System efficiency is a measure of how well a solar energy system is able to convert sunlight into usable electricity. The efficiency is determined by the quality of the solar panels, the inverter, and other components in the system. Solar panels typically have an efficiency rating of around 15-20%, while inverters have an efficiency rating of around 96-98%.

Temperature compensation

The process of adjusting the charging and discharging rates of a battery based on the temperature it is being operated in. Batteries can lose capacity when operated in extreme temperatures, and temperature compensation helps to prolong their lifespan.

Terminal voltage

The voltage of a battery when it is fully charged and not discharging any current. It is the maximum available voltage that a battery can provide.

Thermal runaway

A chemical reaction within a battery that causes it to become overheated and potentially explode. This can occur when a battery is exposed to high temperatures or is overcharged.

Thin-film batteries

Batteries made of lightweight, flexible materials that can be deposited onto a variety of surfaces, including solar panels. They have lower energy density than traditional batteries, but can be useful for certain applications.

Tilt angle

The angle at which solar panels are installed on a roof or ground mount. The tilt angle affects the amount of solar energy that is captured and can be adjusted to optimize energy production based on the location and time of year.

Time-of-use rates

Electricity rates that vary based on the time of day. Solar energy storage systems can be programmed to charge during off-peak hours when rates are lower, and discharge during peak hours when rates are higher.

Top-of-charge voltage

The highest voltage a battery can reach during the charging process. It is important to prevent batteries from being overcharged, as it can cause damage and reduce their lifespan.

Total cost of ownership

The overall cost of a solar energy storage system over its lifetime, including the cost of the system, installation, maintenance, and operation.

Transfer switch

A device that allows for the switching of power sources between the grid and a solar energy storage system. It allows for uninterrupted power supply during a power outage.

Transition charge

The amount of charge remaining in a battery before it transitions into a different stage of charging or discharging. It is important to monitor transition charges to prevent batteries from becoming overcharged or overdischarged.

Ultracapacitor

Ultracapacitors, also known as supercapacitors, are storage devices that store electrical energy through the use of an electrochemical process. Ultracapacitors are capable of storing and releasing energy rapidly and frequently, making them helpful in solar energy storage systems, where energy storage and discharge need to take place in a fraction of a second. Ultracapacitors are also used in electric vehicles and renewable energy systems delivering high power quality and reliability.

Ultrasonic Welding

Ultrasonic welding is a welding process that uses high-frequency vibrations to bond metals and plastic materials. Ultrasonic welding is a high-tech process that ensures a strong and reliable bond between materials at the atomic level. It is a widely used technology in manufacturing battery systems, including solar energy storage batteries. It enables the manufacturer to create a robust and secure battery shell, enhancing the battery's reliability and safety.

Under-voltage Protection

Under-voltage protection is a safety mechanism that prevents a battery's voltage from dropping below safe limits. It is used to prevent the over-discharging of the battery, which can reduce battery life or lead to damage. In solar energy storage systems, under-voltage protection is essential in maintaining the battery's health and enhancing its longevity.

Unified Modelling Language (UML)

Unified modeling language (UML) is a modeling language for software and systems design. It is used to create system diagrams, flowcharts, and other visual models. UML is useful in designing solar energy storage systems by enabling the designer to create a blueprint of the system's structure, relationships, and functionalities. It provides a universal language that helps to communicate project ideas among team members and stakeholders.

Uninterrupted Power Supply (UPS)

An uninterrupted power supply (UPS) is a device that provides backup power by instantaneously switching to battery power in case of a power outage. It is commonly used in residential and commercial settings. In solar energy storage, UPS is used to stabilize the system by providing power when the sun is not shining or when the stored energy in the battery is insufficient. The UPS is also used for managing peak demands on the grid by providing electricity during those periods.

Unit of Energy

A unit of energy is a measure of the amount of energy transferred or transformed. In solar energy storage, the unit of measure used is kilowatt-hours, which measure the amount of energy stored and discharged by the battery system. It is a measure of the energy flow rate over time from the battery into the device or system it is powering. The unit of energy is essential to understand the energy consumption of a solar power system and helps in comparing the energy storage capacity of different solar energy storage systems.

Upgradeability

Upgradeability refers to the capacity of a solar energy storage system to allow easy modification or enhancement of its functionalities, components, or capabilities. In most cases, solar energy storage systems are designed to be upgradeable, which means that their components can easily be modified or expanded as needed. An upgradeable solar energy storage system is cost-effective and efficient in the long term and can adapt to changing energy storage requirements.

User Interface

A user interface (UI) is a point of contact between a human and a machine. In solar energy storage systems, a user interface is used to monitor the battery's charge and discharge cycles, system performance, and fault diagnosis to ensure proper functioning. A good user interface offers a comprehensive and user-friendly way to observe and operate the solar energy storage system.

Utility-Grade Battery Energy Storage System

Utility-grade battery energy storage systems (BESS) are used by power utilities to provide demand management, frequency regulation, load following, and peak shaving. They are large-scale systems that can store and discharge a large amount of electricity. These systems are aimed to maintain the grid's stability and reliability by managing the intermittent solar energy supply. A Utility-grade BESS can store energy from multiple renewable sources such as wind, solar, and hydropower. It requires specialized equipment, and its installation cost is high, but it can be helpful in reducing the carbon footprints of power generation.

Utilization

Utilization refers to the amount of time a solar energy storage system spends discharging its stored energy in relation to the time it has been available. It is a measure of the efficiency of the system and its capability of using its available energy capacity. In solar energy storage, utilization is critical in determining the system's overall efficiency and feasibility of usage.

V2G (Vehicle to Grid)

A system that allows electric vehicles to be used as mobile energy storage units. In a V2G system, the battery of an electric vehicle can be charged with excess solar energy during the day, and then used to power the home or grid during peak energy demand hours.

Vanadium Redox Flow Battery

A type of rechargeable battery that uses vanadium ions in different oxidation states to store and release energy. This type of battery is becoming increasingly popular for solar energy storage, as it offers long-lasting performance and can be easily scaled up or down depending on energy needs.

Virtual Power Plant

A network of distributed energy resources, such as rooftop solar panels, wind turbines, and energy storage systems, that can be controlled centrally to provide grid services and help balance the supply and demand of energy on the grid.

Volt

A unit of measurement for electrical potential energy, representing the amount of energy required to move one coulomb of charge through a circuit.

Voltage

The difference in electrical potential between two points in a circuit. In solar energy storage systems, voltage is used to control the flow of energy between the solar panels and the battery.

Voltage Drop

The loss of electrical potential that occurs when current flows through a circuit. In solar energy storage systems, voltage drop can occur when energy is transported from the solar panels to the battery, and from the battery to the grid.

Voltage Regulation

The process of controlling voltage levels in a solar energy storage system. This is necessary to ensure that the battery is charged and discharged at the right voltage, which helps to maximize energy efficiency and prolong the life of the system.

Voltage-Dependent Charge Current

The rate of current flow in a solar energy storage system, which is determined by the difference in voltage between the batteries and the solar panels. This charge current is essential for replenishing the energy in the battery, and must be carefully regulated to prevent damage to the system.

Volt-Ampere-Hour (VAh)

A unit of measurement for the amount of electrical energy stored in a battery over a specific period of time. This measurement takes into account both the voltage and current of the battery, and is used to determine the overall energy capacity of the system.

VPP Controller

The software or hardware that manages the operation of a virtual power plant. This controller is responsible for balancing the energy supply and demand on the grid, and can also be used to monitor and optimize the performance of individual solar energy storage systems.

Warranty

Many solar energy storage systems come with warranties that vary in length and coverage. Understanding the terms and conditions of the warranty, including what is covered and for how long, is important when choosing a system and for protecting the investment in the system.

Watt (W)

The unit used to measure the power output of a solar panel or storage system. The watt is a measure of energy transfer per unit time, usually measured in kilowatts or megawatts for larger systems. Understanding how much wattage is needed for a desired load is important when designing a solar storage system.

Weight

The weight of a solar storage system depends on its size and components, including the batteries and inverters. Understanding the weight of the system is important when determining installation requirements, such as mounting brackets, and for ensuring the structural integrity of the location where the system will be installed.

Wiring

The electrical connections between solar panels, batteries, and inverters. Proper wiring ensures reliable and efficient energy transfer, while improper or faulty wiring can lead to short circuits and damage to the system. Proper maintenance and inspection of wiring is important for keeping a solar storage system safe and functional.

Yangtze River Delta

The Yangtze River Delta is a vast region in China, including Shanghai and the provinces of Jiangsu, Zhejiang, and Anhui. The region has a high level of solar energy deployment and is a significant solar manufacturing hub. This region is conducive to solar energy storage with its high energy demand, suitable climate, and availability of investment.

Yellow Batteries

Yellow Batteries are rechargeable zinc-air batteries that are free from carbon, lithium, and rare-earth materials. These batteries are long-lived, low cost, and can increase efficiency in storing renewable energy. Yellow batteries store energy in the form of electrochemical reactions that occur when zinc, water, and oxygen react to discharge or recharge.

Yellow Color Code

The yellow color code is used to convey that a cable is used for interconnecting solar modules or panels on an array. The yellow wire is used for the negative connection, while the black wire is used for the positive connection. Yellow cables are UV-resistant, making them ideal for outdoor use in harsh conditions.

Yelomine

Yelomine is a PVC-based pipe and fitting system used in solar-powered water systems. It is corrosion-resistant and UV-protected, making it ideal for use in harsh environments. Yelomine is a durable, cost-effective alternative to metal pipes and fittings for solar-powered water systems. Yelomine pipe fittings are available in various sizes to accommodate various pipe diameters.

Yield

The amount of power generated by a solar panel is its yield. Measured in watts or kilowatts, yield depends on various variable factors like panel orientation, temperature, shading, and time of day. A solar panel's yield varies across countries and regions, depending on the amount of sunlight received in a specific area.

Yield Guarantee

Yield guarantee is a type of insurance or warranty that provides assurance regarding a specific solar panel, array, or system's energy output. This guarantee gives users confidence in the performance of the solar system, and companies offer yield guarantees to attract customers. Yield guarantees stipulate that if the system doesn't perform as promised, the company must either repair, replace, or refund the customer.

Yield Optimization

Yield optimization refers to the method of maximizing the power output from a solar array. It involves factors such as the angle of the solar panels, the arrangement of the array, and the geographic location of the array. Yield optimization ensures that the solar array operates at maximum efficiency, resulting in more energy generation and greater ROI for investors.

Yieldco

A Yieldco is a publicly traded company that develops, owns, and operates renewable energy projects. Yieldcos own and manage large-scale solar or wind farms, selling the energy generated to utilities or other customers. Yieldcos offer a stable investment option for retail investors, with dividends paid from the revenue generated through the sale of electricity. Yieldcos also provide project developers and sponsors with a source of long-term financing.

Ylem Energy

Ylem Energy is a UK-based specialist in energy efficiency and renewable energy. Ylem Energy emphasizes the importance of integrating energy storage with renewable energy generation. Its clients include industrial and commercial companies, as well as major organizations in the public sector. The company takes a consultative approach, working with clients to develop tailored solutions to their specific energy needs.

Yotta Energy

Yotta Energy is a US-based company that specializes in energy storage systems. It has developed a proprietary energy storage technology that uses a unique chemical reaction to produce long-lasting and low-cost energy storage. Yotta Energy's modules are scalable and can be tailored to meet specific customer needs. Its technology has applications in solar energy storage, grid stabilization, and backup power.